AF575858

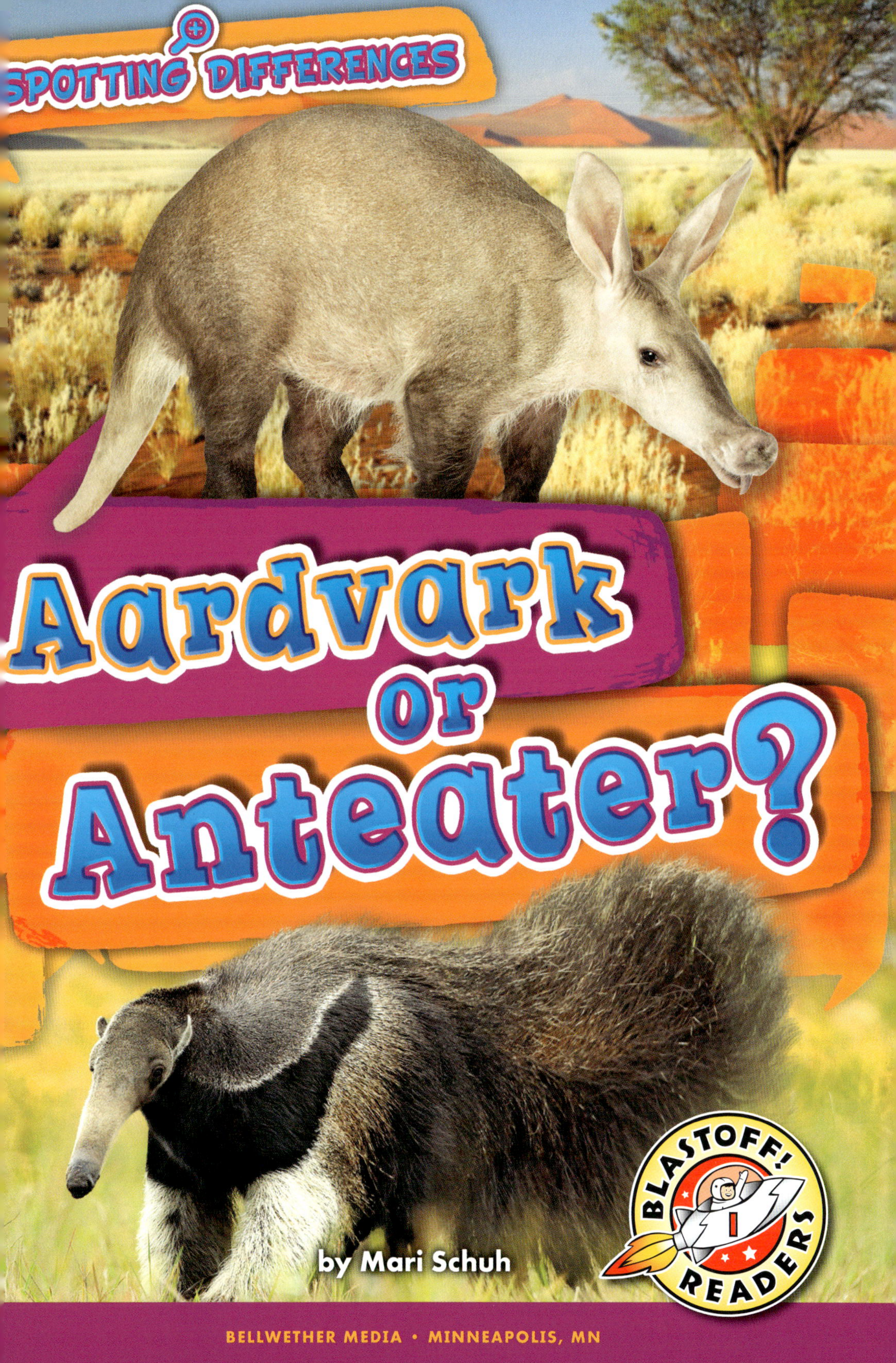
SPOTTING DIFFERENCES
Aardvark
or
Anteater?
by Mari Schuh
BLASTOFF! READERS
1
BELLWETHER MEDIA • MINNEAPOLIS, MN

Blastoff! Readers are carefully developed by literacy experts to build reading stamina and move students toward fluency by combining standards-based content with developmentally appropriate text.

Level 1 provides the most support through repetition of high-frequency words, light text, predictable sentence patterns, and strong visual support.

Level 2 offers early readers a bit more challenge through varied sentences, increased text load, and text-supportive special features.

Level 3 advances early-fluent readers toward fluency through increased text load, less reliance on photos, advancing concepts, longer sentences, and more complex special features.

★ **Blastoff! Universe**

Reading Level

Grade
K

Grades
1–3

Grade
4

This edition first published in 2023 by Bellwether Media, Inc.

Library of Congress Cataloging-in-Publication Data

Names: Schuh, Mari C., 1975- author.
Title: Aardvark or anteater? / by Mari Schuh.
Description: Minneapolis, MN : Bellwether Media, Inc., 2023. | Series: Blastoff! readers: spotting differences | Includes bibliographical references and index. | Audience: Ages 5-8 | Audience: Grades K-1 | Summary: "Developed by literacy experts for students in kindergarten through grade three, this book introduces aardvarks and anteaters to young readers through leveled text and related photos"-- Provided by publisher.
Identifiers: LCCN 2021062907 (print) | LCCN 2021062908 (ebook) | ISBN 9781644876954 (library binding) | ISBN 9781648347412 (ebook)
Subjects: LCSH: Aardvark--Juvenile literature. | Myrmecophaga--Juvenile literature.
Classification: LCC QL737.T8 S38 2023 (print) | LCC QL737.T8 (ebook) | DDC 599.3/1--dc23/eng/20220104
LC record available at https://lccn.loc.gov/2021062907
LC ebook record available at https://lccn.loc.gov/2021062908

Editor: Elizabeth Neuenfeldt   Designer: Laura Sowers

Printed in the United States of America, North Mankato, MN.

Table of
Contents

Aardvarks and Anteaters

Aardvarks and anteaters are **mammals**. They both have long **snouts**.

anteater
snout

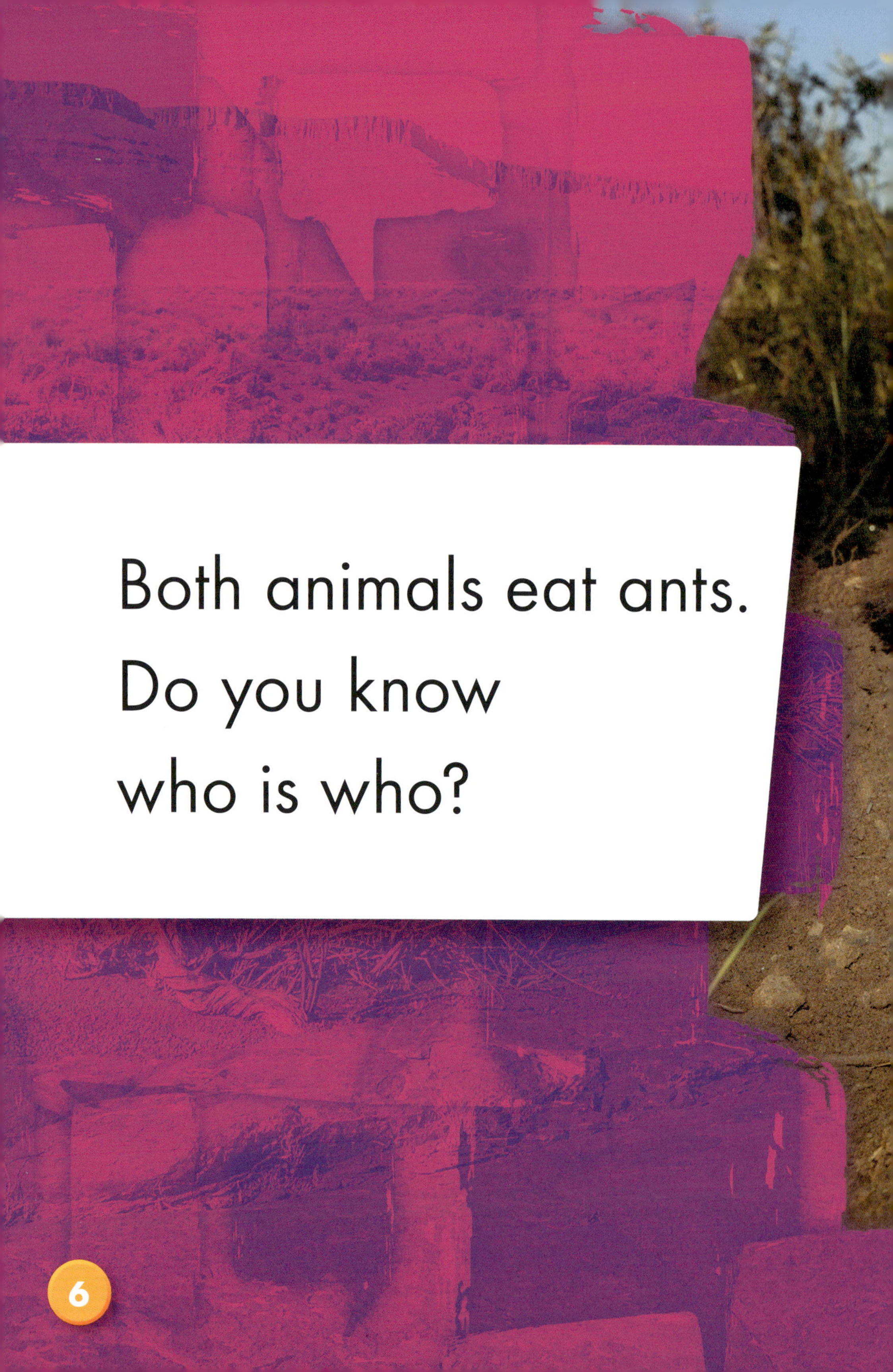

Both animals eat ants.
Do you know
who is who?

aardvark
ants

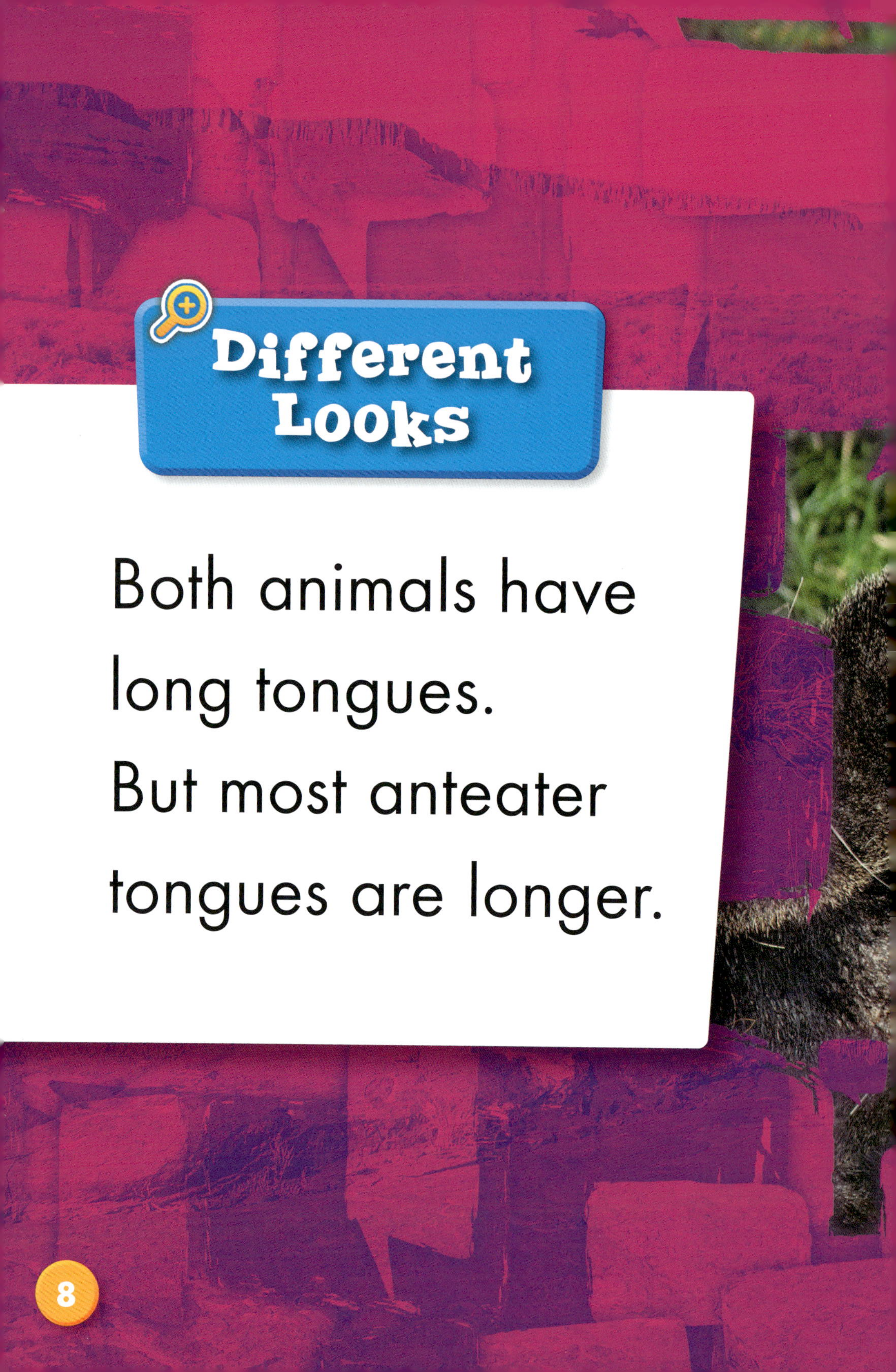

Different Looks

Both animals have long tongues. But most anteater tongues are longer.

tongue

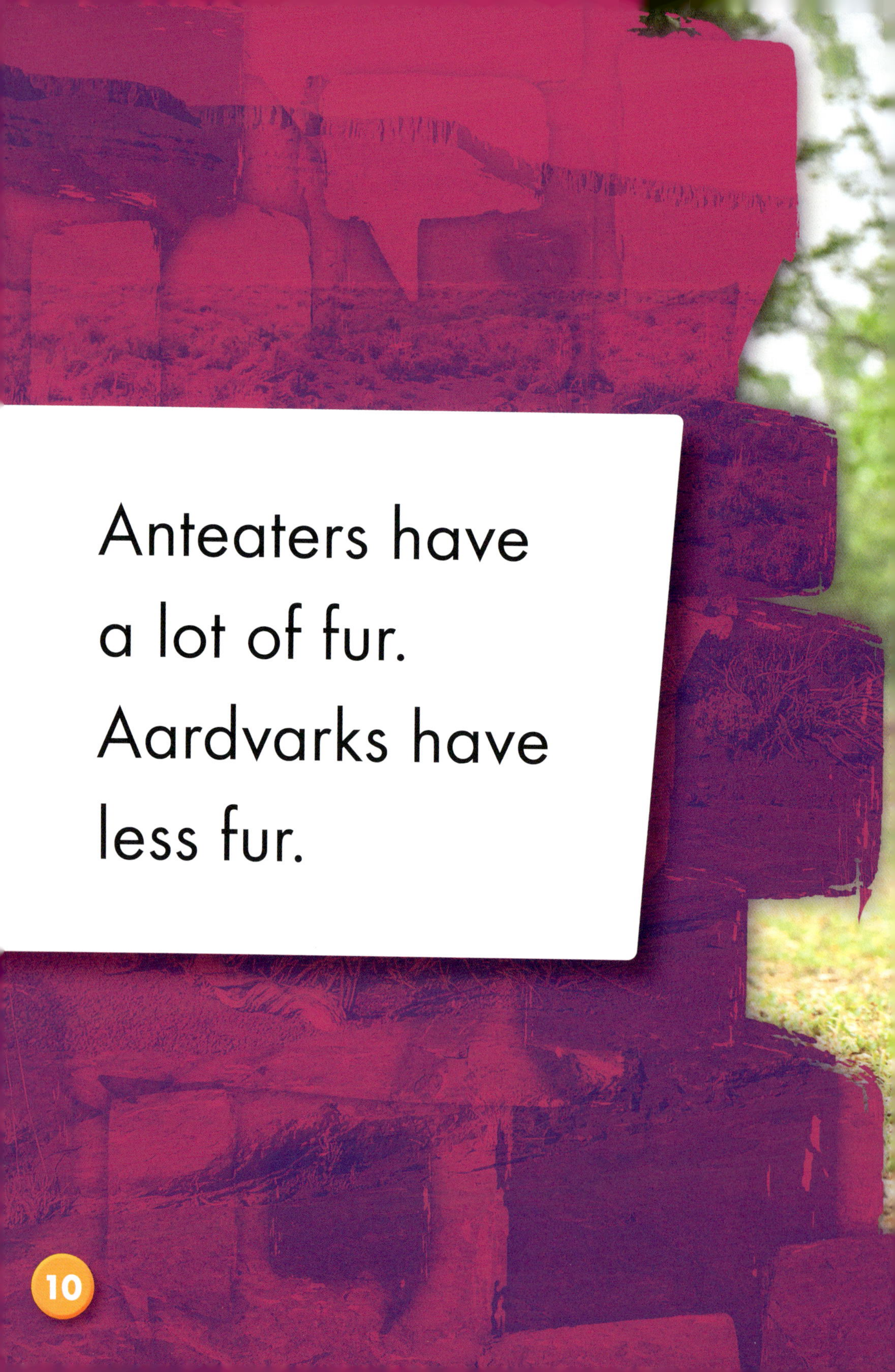

Anteaters have a lot of fur. Aardvarks have less fur.

Aardvarks have big ears. Anteaters have small ears.

ear

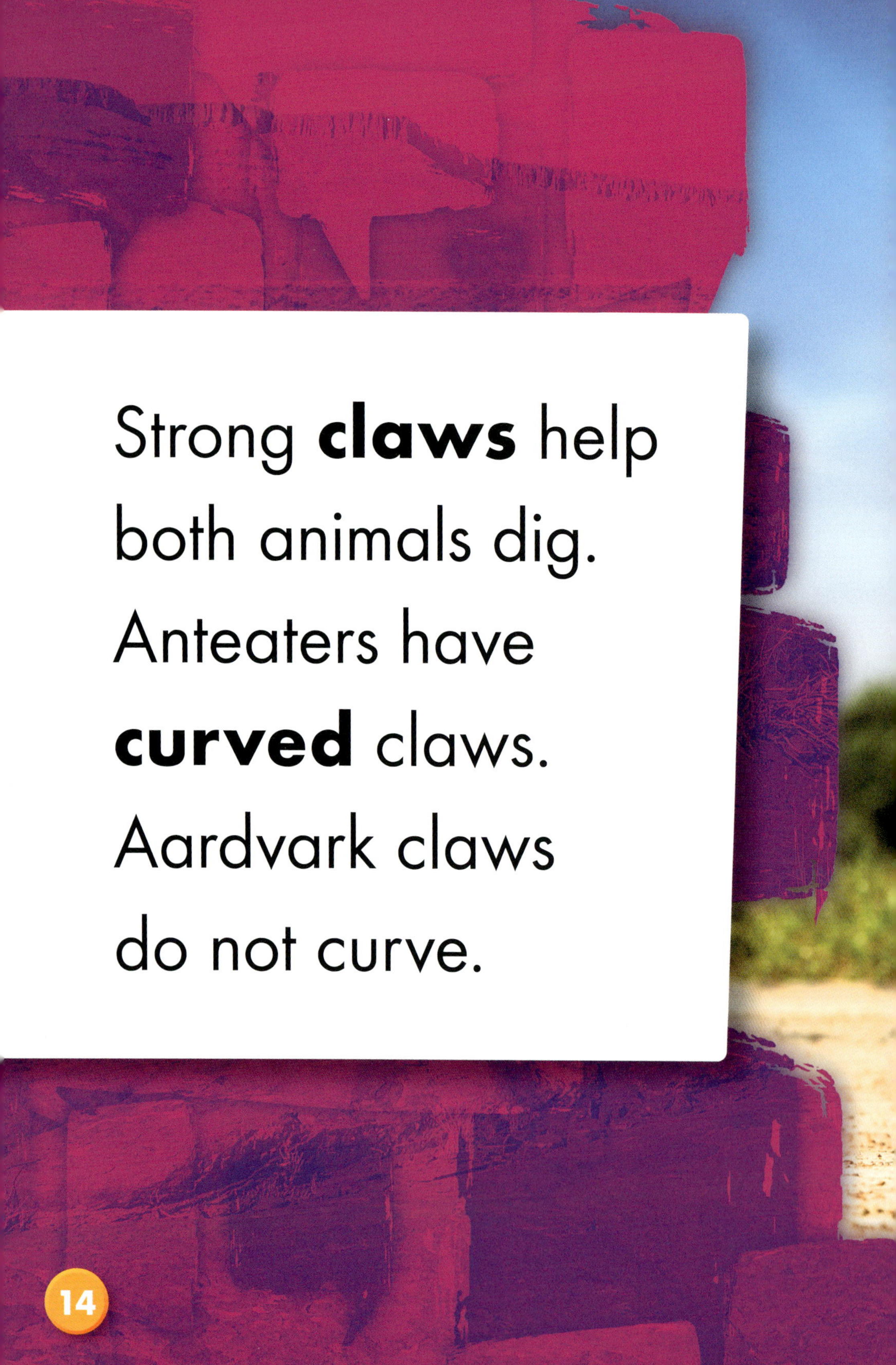

Strong **claws** help both animals dig. Anteaters have **curved** claws. Aardvark claws do not curve.

curved
claws

Different Lives

These animals live far apart. Anteaters live in South America and **Central America**. Aardvarks live in Africa.

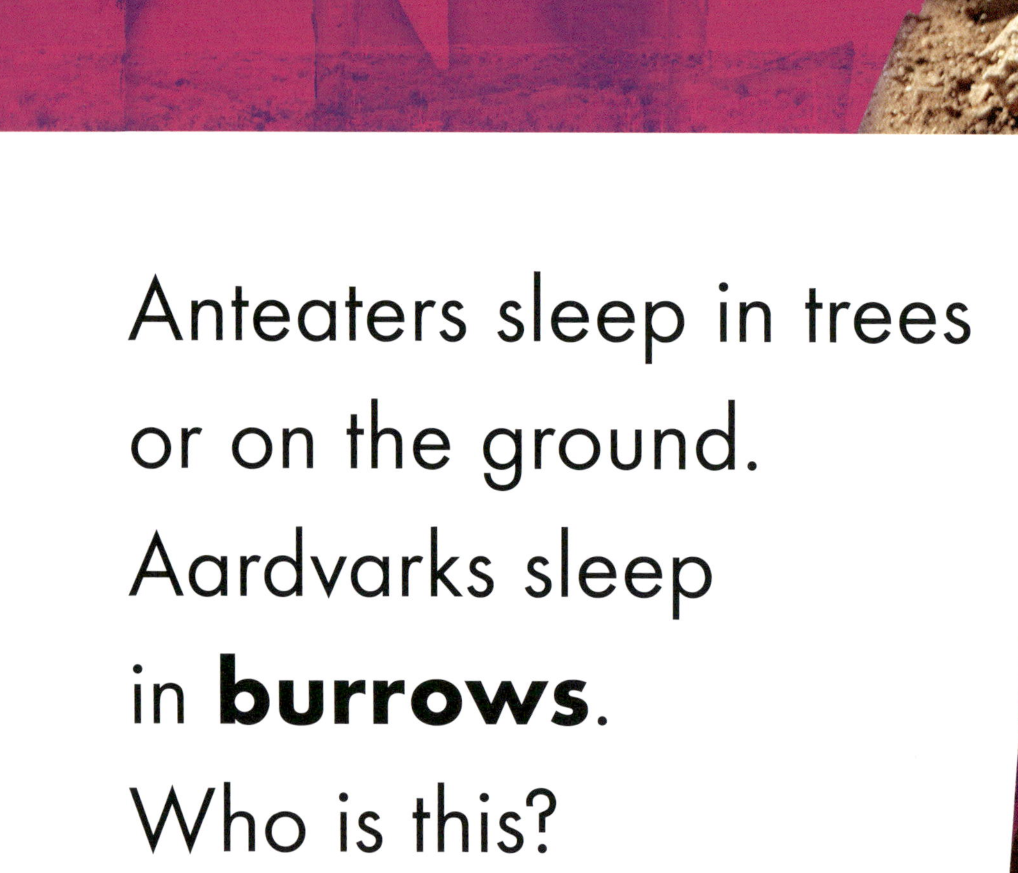

Anteaters sleep in trees
or on the ground.
Aardvarks sleep
in **burrows**.
Who is this?

Side by Side

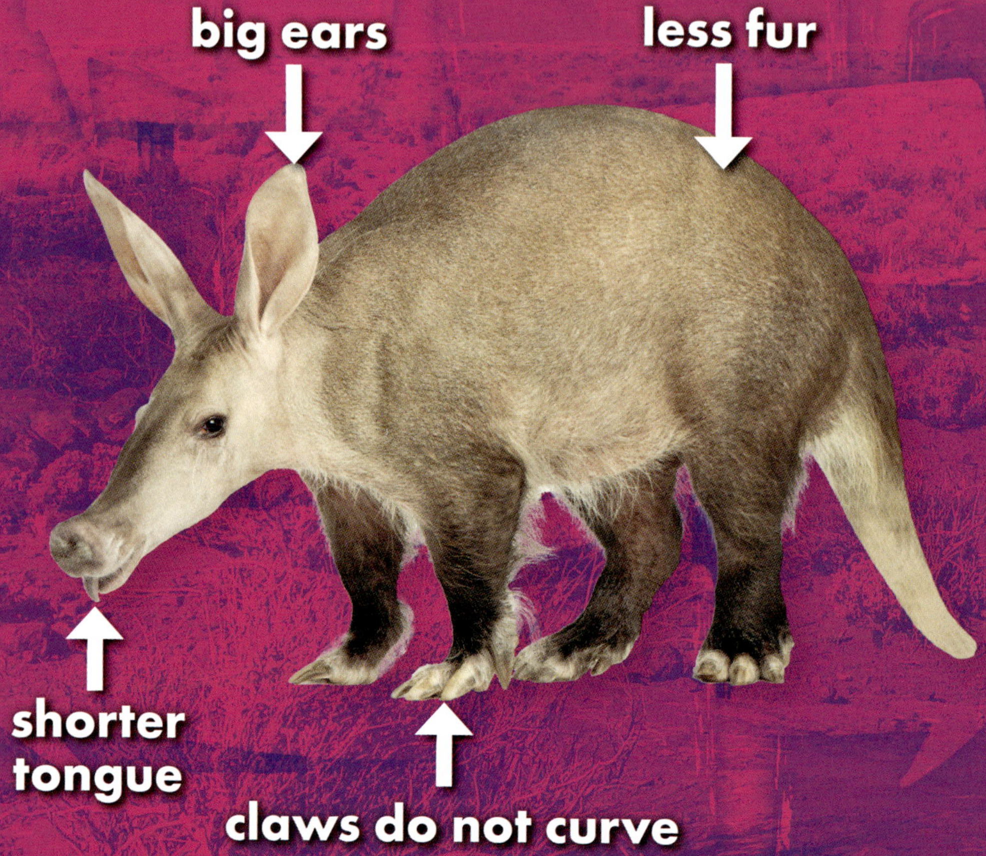
big ears
less fur
shorter tongue
claws do not curve

Aardvark Differences

live in Africa

sleep in burrows

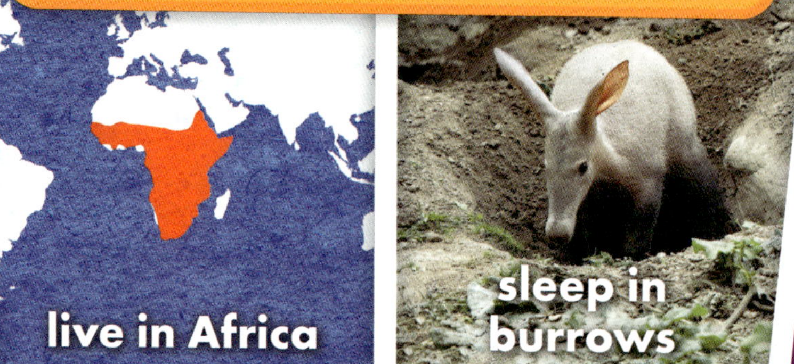

a lot of fur
small ears
longer tongue
curved claws
Anteater Differences
live in Central and South America
sleep on the ground or in trees

Glossary

burrows

holes or tunnels in the ground made or used by animals

curved

having a bend

Central America

the countries between Mexico and South America

mammals

warm-blooded animals that have backbones and feed their young milk

claws

sharp nails on the toes of animals

snouts

the long noses and mouths of some animals

To Learn More

AT THE LIBRARY

Bassier, Emma. *Aardvarks.* Minneapolis, Minn.: Pop!, 2020.

Meister, Cari. *Do You Really Want to Meet an Anteater?* Mankato, Minn.: Amicus Ink, 2019.

Orr, Tamra B. *Aardvark or Anteater.* Ann Arbor, Mich.: Cherry Lake Publishing, 2020.

ON THE WEB

FACTSURFER

Factsurfer.com gives you a safe, fun way to find more information.

1. Go to www.factsurfer.com.
2. Enter "aardvark or anteater" into the search box and click 🔍.
3. Select your book cover to see a list of related content.

Index

The images in this book are reproduced through the courtesy of: Eric Isselee, front cover (aardvark), p. 20 (aardvark); coPrint, front cover (top background); Ondrej Prosicky, front cover (anteater), pp. 4-5, 13, 17, 22 (snouts); Avalon.red/ Alamy, pp. 6-7; MBL1, p. 7 (ants); esdeem, pp. 8-9; Rico Leffanta, pp. 9, 11; Braian Moreno, pp. 10-11; Geno EJ Sajko Photography, pp. 12-13; Nature Picture Library/ Alamy, pp. 14-15, 22 (curved); Biosphoto/ Alamy, p. 15; Afripics/ Alamy, pp. 16-17; S. Tuengler - inafrica.de/ Alamy, pp. 18-19; Vladislav T. Jirousek p. 20 (sleep in burrows); gan chaonan, p. 21 (anteater); Rabea Hirschfeld, p. 21 (sleep on ground); Dietmar Rauscher, p. 22 (burrows); NNehring, p. 22 (claws); PhotocechCZ, p. 22 (mammals).